Overheard at the Square Dance

Edited by
Gordon Goss & Valerie Thornton

A Citadel Press Book
Published by Carol Communications

Previously published as *America Speaks*.

Copyright © 1988 by Gordon Goss

A Citadel Press Book
Published by Carol Communications

Editorial Offices
600 Madison Avenue
New York, NY 10022

Sales & Distribution Offices
120 Enterprise Avenue
Secaucus, NJ 07094

In Canada: Musson Book Company
A division of General Publishing Co. Limited
Don Mills, Ontario

Queries regarding rights and permissions
should be addressed to: Carol Communications,
600 Madison Avenue, New York, NY 10022

Manufactured in the United States of America
ISBN 0-8065-1120-6

ON THE HUMOROUS SIDE

1

What do you mean I'm overdrawn? I still have checks left!

2

Why can't life's big problems come when we are teenagers and know everything?

3

Never judge a book by the movie with the same title.

4

Remember when we would talk out our problems over coffee and a cigarette? Now they are the problems.

5

If you can read the handwriting on the wall, your children are old enough to know better.

6

The next time you see a dollar on the street, pick it up. There might be something valuable underneath it.

7

My decision is maybe and that's final.

8

The modern wife is one who knows what her husband's favorite dishes are - and the restaurants that serve them.

9

Did you ever wonder how many fig leaves Eve tried on before she said, "I'll take this one"?

10

The optimist sees the donut, the pessimist sees the hole, and the realist sees the calories.

11

I was winning the game of life, but there was a flag on the play.

12

The loudest noise known to man is the first rattle in a brand new car.

13

Crime wouldn't pay if the government ran it.

14

An intellectual is someone who can listen to the William Tell Overture without thinking of the Lone Ranger.

15

One thing about the speed of light - it gets here too early in the morning.

16

When you come to the end of a perfect day, it probably isn't over yet.

17

The severity of an itch is inversely proportional to the reach.

18

A bachelor is a guy who can make a fool out of himself without being told about it.

19

I may rise, but I refuse to shine.

20

People who carved out a place in history didn't do it by chiseling.

21

Just because you're paranoid doesn't mean someone isn't after you.

22

Most teenager's aptitude tests find them best suited for retirement.

23

If you kicked the one responsible for most of your troubles, you wouldn't be able to sit down for six months.

24

Life is like an onion, you peel off one layer at a time and sometimes you weep.

25

A flood is a river that's gotten too big for its bridges.

26

My favorite time to get up is the crack of noon.

27

To error is human, but to really screw things up you need a computer.

28

The best two people in the world are you and me, but sometimes I wonder about you.

29

People who live in glass houses shouldn't run around naked.

30

God, please give me patience and hurry up about it.

31

The only things in life that are certain are death, taxes and holes in your socks.

32

Life may be a bowl of cherries, but I only get the pits.

33

If you give me a home where the buffalo roam, please clean up after them.

34

If "ifs" and "buts" were candy and nuts, we'd all have fruitcake for Christmas.

35

If I want your opinion, I'll give it to you.

36

Almost only counts in horseshoes and hand grenades.

37

In the United States a woman has a baby every 20 seconds. To control our population, we must find this woman and stop her.

38

I know I'm okay because God don't make junk!

39

When you see a light at the end of the tunnel, hopefully it won't be a freight train coming towards you.

40

Some men will allow women to make fools of them; others are the do-it-yourself type.

41

Birthdays are like jogging shorts . . . they creep up on you.

42

People who think they know it all annoy those of us who already do.

43

Three may keep a secret if two of them are dead.

44

The road to success is always under construction.

45

The world is full of willing people, some willing to work, the rest willing to let them.

46

A seefood diet is the best - whenever you see food, eat it.

47

Computers that check your spelling are great, but I need one that checks my thinking.

48

Daddy, you don't have to pay for it, just charge it!

49

Why do I always have so much month left after the end of my paycheck?

50

Let us toast the fools because without them the rest of us could not succeed.

51

My "get up and go" got up and went.

52

Read a good book regularly, even at the risk of straining your mind.

53

A girdle is a device used to keep an unfortunate condition from spreading.

54

A bachelor is a fellow who failed to embrace his opportunities.

55

Big shots are only little shots who keep shooting.

56

If you can smile when everything goes wrong - you must be a repairman.

57

The strong take it away from the weak, the clever take it away from the strong, and the government takes it away from everybody.

58

A class reunion is where everyone gets together to see who is falling apart.

59

Don't ever doubt your wife's judgement - look who she married!

60

The loudest boos always come from those in the free seats.

61

If all the people who go to sleep in church were laid end to end, they would be a lot more comfortable.

62

The difference between being diplomatic and undiplomatic is the difference between saying "When I look at you, times stand still," and saying "Your face would stop a clock."

63

Remember when "enter" was a sign on a door and not a button on a computer?

64

A man has reached middle age when he doesn't care where his wife goes so long as she doesn't make him come along.

65

Smuggling perfume is a fragrant violation of the law.

66

Cheer up! Birds also have bills, but they keep on singing.

67

The automobile manufacturers' next big problem is how to get their front-wheel drives to listen to a backseat driver.

68

We would all be happy to pay as we go - if only we could finish paying for where we've been.

69

A jury consists of twelve people chosen to decide who has the better lawyer.

70

Remember when a compact was the ideal gift for a teenage girl? Well, it still is - providing it has four wheels.

71

Most people don't mind suffering in silence as long as everyone else knows about it.

72

The world is divided into two kinds of people, the good and the bad - and the good decide which is which.

73

If you want to see a short summer, borrow some money due in the fall.

74

If you tell people there are a billion stars in the universe, they'll believe you. But if you tell them a bench has just been painted, they have to touch it to be sure.

75

People can be divided into three groups; those who make things happen, those who watch things happen, and those who wonder what happened.

76

Hibernation is a bear necessity.

77

A penny saved is not worth much today.

78

Life insurance is a bet that you will die.

79

Old bridge players never die, they just "PASS" away.

80

The early bird cooks his own breakfast.

81

There ought to be a better way to start the week than Monday.

82

If God had wanted us to see the sunrise, He would have made it rise at noon.

83

When I think I can make ends meet, somebody moves the ends.

84

Behind every successful man is the IRS.

85

An apple a day keeps the doctor away, but an onion a day keeps everybody away.

86

Everybody tells you to have a nice day but nobody tells you how.

87

Those people who think of themselves as wits are often half right.

88

What I'd most like to get out of my new car is my teenager.

89

Discovery of a new food dish gets man more excited than a great scientific discovery.

90

An expert is one who knows more and more about less and less.

91

If someone says it's the principle of the thing and not the money, you can bet it's the money.

92

These days it almost takes more brains to fill out your income tax than it does to make the money.

93

If opportunity knocks on your door, by the time you unlatch the bolt, turn the deadbolts, unlock the chain and silence the burglar alarm, it will be gone.

94

Tact is the ability to close your mouth before someone else wants to.

95

After all is said and done, more is said than done.

96

The most disappointed people in the world are those who get what is coming to them.

97

No one gets too old to learn a new way of doing something dumb.

98

Some people who do their duty as they see it may be in the need of glasses.

99

Don't worry if you start losing your memory. Just forget about it.

100

Woman's work that's never done is most likely what she asked her husband to do.

101

Ants don't go to picnics - we take picnics to them.

102

If exercise does one so much good, how come athletes have to retire at 35?

103

A vacation is two weeks spent with those you thought you loved the best.

104

A baker is a person who is always rolling in dough.

105

Opportunity knocks; temptation kicks the door down.

106

Be tolerant of those who disagree with you - after all, they have a right to their ridiculous opinion.

107

The trouble with being a leader is that you don't know whether the people are following you or chasing you.

108

Never serve meals on time. The starving eat anything.

109

Some people learn traffic rules by accident.

110

Grandchildren of any age can always make their grandparents happy just by saying, "I'm hungry!"

111

He who has all the answers hasn't heard all the questions.

112

There isn't much to see in a small town, but what you hear makes up for it.

113

It is better to cause happiness where you go than to cause happiness when you go.

114

It's better to tighten your belt than lose your pants.

115

An icicle is a drip caught in a draft.

116

A woman is as young as she feels like telling you she is.

117

An optimistic gardener is a person who believes that what goes down must come up.

118

A gossip usually gets caught in their own mouth-trap.

119

Marriages are made in heaven - so are thunder and lightening.

120

A fool and his money are soon parted, the rest of us wait for tax time.

121

Why is it when you dial a wrong number, you never get a busy signal.

122

If your palm itches, it's a sign that your going to get something. If your head itches, you've got it.

123

He always wore a rainbow tie because there was a big pot at the end of it.

124

A wallet is a device that permits you to lose all your valuables at the same time.

125

Those proud of keeping an orderly desk never know the thrill of finding something you thought you had irretrivably lost.

126

Happiness is the 15th day of a 14-day diet.

127

A diplomat is anyone who thinks twice before saying nothing.

128

A dog has lots of friends and fun maybe because he wags his tail and not his tongue.

129

Egotism is that certain something that enables a man in a rut to think he's in the groove.

130

The difference between gossip and news is whether you hear it or tell it.

131

Vacations are wonderful; they make you feel good enough to go back to work and poor enough so that you have to.

132

Why is it that an empty guest room always has full closets?

133

They say children brighten up the home. That's right - they never turn off the lights.

134

The way they are always full, a parking lot should be called a parking little.

135

After paying for the wedding, about all a father has left to give away is the bride.

136

If you think old soldiers just fade away, try getting into your old Army uniform.

137

People who think nothing is impossible have never tried to settle an account with a computer.

138

We shouldn't feel too bad, prehistoric man was never able to anticipate a drop in the rock market.

139

When the going gets tough, the tough go shopping.

140

How could I be "over the hill" if I was never at the top of it.

141

A bargain is something you can't use at a price you can't resist.

142

It's not as easy as you think to get a parking ticket. First you must find a parking space.

143

Golf is the game that took the cow out of the pasture and let the bull in.

144

I never hold a grudge after I get even.

145

If you want an answer, see the boss; but if you want the correct answer see his secretary.

146

People who snore always fall asleep first.

147

An optimist is the man who married his secretary, thinking he'd be able to go on dictating to her.

148

If we are what we eat, nuts must be a common diet.

149

Life is like a shower, one wrong turn and you are in hot water.

150

To err is human; to blame it on someone else is even more human.

151

I shop like a bull - I charge everything.

152

Owning an ant farm does not qualify you as a group leader.

153

An owl with laryngitis doesn't give a hoot.

154

We're all self-made but only the rich will admit it!

155

A compromise is an arrangement whereby people who can't get what they want make sure nobody else does, either.

156

The things people want to know about are usually none of their business.

157

We'll never really crack down on air pollution until it interferes with our TV reception.

158

Disneyland is the only people trap ever built by a mouse.

159

Be Alert. Lerts live longer.

160

Work is the greatest thing in the world so we should always save some of it for tomorrow.

161

Things always get better or worse, so we're either worrying for nothing or too soon.

162

Show me a man with head held high and I'll show you a man who can't get used to bifocals.

163

The optimist proclaims that we live in the best of all possible worlds; and the pessimist fears this is true.

164

The trouble with good advice is that it usually interferes with our plans.

165

Whenever your arms are filled with packages, the sign on the door says "Pull".

166

If at first you don't succeed, try looking in the wastebasket for the instructions.

167

My hair has had three basic styles: parted, unparted and departed.

168

Old age is 20 years older than you are.

169

I'm on a pasta diet - my doctor says I've been off my noodle long enough.

BEFORE　　　　**AFTER**

170

The Iran-Contra fiasco shows that one can achieve immortality through spectacular error.

171

When a person goes on a diet the first thing he loses is his temper.

172

Rain is caused by high-pressure areas, cold fronts, warm moist air, a newly washed car and the first day of your vacation.

173

America is the only country in the world where men get together to talk about hard times over a $15.00 steak.

174

Confirmed bachelors, like detergents, work fast and leave no rings.

175

Some people have a head like a door knob - anybody can turn it.

176

You've reached middle age when your wife tells you to pull in your stomach and you already have.

177

The older a man gets, the farther he had to walk to school as a boy.

178

If at first you don't succeed, you'll get a lot of advice.

179

If all else fails, go shopping.

180

Talk is cheap because the supply exceeds the demand.

181

Most of us carry our own stumbling block with us - we camouflage it with a hat.

182

Management is a series of interruptions, interrupted by interruptions.

183

Diet is the penalty for exceeding the feed limit.

184

Another thing man can do that the lower animals can't is to stand upright in front of a crowd and put both feet in his mouth.

185

The only time that SUCCESS comes before WORK is in the dictionary.

186

Attics are better than garages. You don't have to keep cleaning them out to make room for the car.

187

A bargin hunter is a person moved by the spirit of brotherly shove.

188

The "Iron Age" ended with permanent press.

189

Few things are harder to put up with than the annoyance of a good example.

190

Our strength as humans is that we can laugh at ourselves for being ridiculous. Our weakness is that we have to do it so often.

191

You are getting old if the exercise you use to do for a warm-up is now a stiff work-out.

192

Stand up to be seen, speak up to be heard, shut up to be appreciated.

193

Experience is a wonderful thing. It enables you to recognize a mistake when you make it again.

194

Plastic surgeons can do almost anything with a nose, except keep it out of other people's business.

195

An optimist is someone who tells you to cheer up when things are going his way.

196

A skunk is a streamlined cat with a two-toned finish and a fluid drive.

197

Maybe the old-time doctor didn't know what was wrong with you, but he didn't charge you $25 to send you to somebody who did.

198

One reason you can't take it with you is that it goes before you do.

199

Middle age spread brings people closer together.

200

At least when you talk to children, they don't show you pictures of their parents.

201

People with laryngitis sure don't complain much.

202

Common sense isn't very common these days.

203

People who boast about their family tree have usually pruned it first.

204

How come when you go to bed at night the room is warm and the bed is cold; but when you wake up in the morning, the bed is warm and the room is cold?

205

An ant is a small insect which is always at work but still finds time to go to picnics.

206

Few persons ever grow up; they merely change their playthings.

207

Mealtime is when the kids sit down to continue eating.

208

Hay fever can be either positive or negative; sometimes the eyes have it and sometimes the nose.

209

An alarm clock is a mechanism used to scare the daylights into you.

210

A race track is the only place to find windows that clean people.

211

You're getting old when the gleam in your eye is from the sun hitting your bifocals.

212

A pessimist is someone who likes to listen to the patter of little defeats.

213

If at first you don't succeed, you're running about average.

214

Have you ever noticed that the perforated part of a postage stamp is stronger than the solid part?

215

Money may not bring happiness but it would be nice to find out for yourself.

216

Living in the past has one thing in its favor - it's cheaper.

217

A traffic light is a little green light that changes to red as your car approaches.

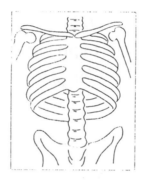

218

Never argue with your doctor - he has inside information.

219

Don't smoke in bed - the ashes that fall on the floor may be your own.

220

Nowadays they spend $30,000 for a school bus to pick the kids up right at the door; then they spend $30,000 for a gym so they can get some exercise.

221

Garlic is highly recommended for colds. Eat it and everybody who has a cold will stay away.

222

He who indulges, bulges.

223

Nowadays a man who is contented with his lot probably has a house on it.

224

The trouble with being a good sport is that you have to lose in order to prove it.

225

If you look like your passport photo, you need the trip.

226

The promise of some people to be on time carries a lot of wait.

227

Be friendly with the folks you know; if it weren't for them, you'd be a total stranger

228

Keep your temper! No one else wants it.

229

If athletes get athlete's foot, do astronauts get mistletoe?

230

Even more ominous than the sound of a riot is a bunch of kids suddenly quiet.

231

By the time our children are old enough not to do anything in public to disgrace us, they have reached an age when the things we do embarrass them.

232

In the old days the man who saved money was a miser; nowadays he's a miracle worker.

233

A procrastinator is one who puts off until tomorrow the things he's already put off until today.

234

All you have to do to get the world to beat a path to your door is decide that you want to take a nap.

235

Worry is like a rocking chair . . . it gives you something to do but doesn't get you anywhere.

236

The noblest of all dogs is the hot dog; it feeds the hand that bites it.

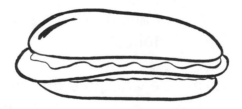

237

Never go to a doctor whose office plants have died.

238

Oh, Lord, make my words palatable because tomorrow I may have to eat them.

239

When I work, I work hard; when I sit, I sit loose; and when I think I fall asleep.

240

A closed mouth gathers no feet.

241

An expert is one who avoids small errors as he sweeps to his final mistake.

242

The three-day weekend was created because it's impossible to cram all the bad weather into two days.

243

In today's supermarkets, the shopping carts easily attain speeds of more than $55 per hour.

244

Never argue with a fool for onlookers might not be able to tell which is which.

245

When the grass looks greener on the other side of the fence, it may be that they take better care of it.

246

It's better for things to go in one ear and out the other than to go in one ear, get all mixed up, and then slip out of the mouth.

247

If my ship ever does come in, I'll probably be waiting at the airport.

248

You know you've reached middle age when your knees buckle and your belt doesn't.

249

A garage sale is a technique for distributing all the junk in your garage among all the other garages in the neighborhood.

250

One of the greatest mysteries of life is how the boy who was not good enough to marry your daughter can be the father of the smartest grand child in the world.

251

A great oak is only a little nut that held his ground.

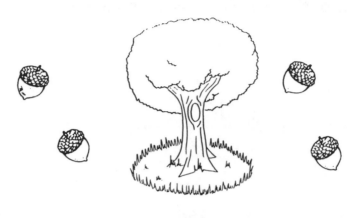

252

The perfect host is one who makes his guests feel at home even though he wishes they were.

253

If Patrick Henry thought that taxation without representation was bad, he should see how bad it is with representation.

254

If you think you're lonely and neglected just think of Whistler's father.

255

Humility is the ability to look properly shy while you tell people how wonderful you are.

256

Two words that should never be used together - Army Intelligence.

257

Unfortunately when you meet a girl that is a perfect "10", she usually has an I.Q. to match it.

258

The trouble with advice is that you seldom know whether it's good or bad until you no longer need it.

259

People with the narrowest minds always seem to have the widest mouths.

260

Even a woodpecker owes his success to the fact that he uses his head.

261

We must admit that the younger generation learns fast, but we are not sure what.

262

Winter is the time of the year when it's too cold to do the jobs around the house that didn't get done in the summer because it was so hot.

263

A screen door is something the grandkids get a bang out of.

264

A prescription is hieroglyphics written by a physician to be translated by a pharmacist into dollar signs.

265

The average taxpayer may be America's first natural resource to be exhausted.

266

He took his defeat like a man; he blamed it on his wife.

267

No matter how busy you are, you are never too busy to stop and talk about how busy you are!

268

The only way you can make a living harping at others is to play in an orchestra.

269

Good writing, like good cooking, can often use a little shortening.

270

What melts in your mouth will bulge in your hips.

271

How come parents say they have a new baby? Who ever heard of an old baby?

272

Inflation is just a drop in the buck.

273

A wedding ring is a one-man band.

274

A gourmet avoids unfashionable restaurants because he doesn't want to gain weight in the wrong places.

275

The mind is a scheme engine.

276

Among the things so simple a child can operate are parents.

277

An egotist is one who suffers from "I" strain.

278

You can't blame parents if they cry a little at their daughter's wedding. Weddings are mighty expensive.

279

Overeating will make you thick at your stomach.

280

Strange that people call money "dough". Dough sticks to your fingers.

281

Firmness is that admirable quality in ourselves that is merely stubbornness in others.

282

A good friend is one who can keep a straight face while helping you repair your latest "do-it-yourself" project.

283

Most people would rather defend to death your right to say it than listen to it.

284

Our population is made up of the haves, the have nots, and the haven't paid for it yets.

285

Some people are so busy learning the tricks of the trade they never learn the trade.

286

There is no guaranteed way to lose weight, but living on Social Security comes close to it.

287

The last accurate weather forecast just may have been when God told Noah that it was going to rain.

288

Money may not be everything, but it is a great consulation until you have everything.

289

Ten cents used to be big money, but dimes have changed.

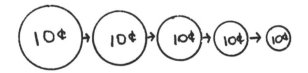

290

You should teach a child to wait on himself - even if you are married to that child!

291

Convenience food is anything that's at the front of the refrigerator.

292

People who can't open child-proof bottles are adults.

293

If you have trouble going to sleep at night, lie at the very edge of the bed - you'll soon drop off.

294

Wasn't life better when you had ten speeds and bikes didn't?

295

If bankers can count, how come they have eight windows and only four tellers?

296

What it takes to do the job of three men: one woman.

297

Automatic transmissions were made for people too lazy to shift for themselves.

298

Today is the day to make firm decisions - or is it?

299

If the shoe fits, it's the wrong color.

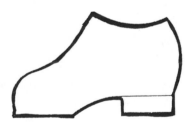

300

A safety belt is the best way to keep from leaving the scene of an accident.

301

The telephone is a remarkable invention. It allows teenagers to go steady without holding hands.

302

Diplomacy is telling your husband he has an open mind, not holes in his head.

303

Credit cards are buy passes.

304

In December we jingle bells; in January we juggle bills.

ON THE
SERIOUS SIDE

305

Minds are like parachutes - they only function when open.

306

Wrinkles merely indicate where many smiles have been.

307

When your work speaks for itself, don't interrupt.

308

The secret of success is to start from scratch and keep on scratching.

309

Health is the most unappreciated blessing in the world - by those who have it.

310

It is always easier to hate something than it is to understand it.

311

The squeaking wheel doesn't always get the grease; sometimes it gets replaced.

312

At the age of twenty we don't care what the world thinks of us; at thirty we worry about what it thinks of us; at forty we discover that it isn't thinking of us at all.

313

You have to learn from the mistakes of others; you can't possibly live long enough to make them all yourself.

314

Temper is what gets most of us into trouble; pride is what keeps us there.

315

A friend is one who knows all about you and still likes you.

316

If seed in the black earth can turn into such beautiful flowers, what might the heart of man become in its long journey toward the stars?

317

Having a good friend is never being lonely.

318

A sense of curiosity is nature's original school of education.

319

Friendship is the bloom in life's garden.

320

A smile is an inexpensive way to improve your looks.

321

Freedom is not a question of doing as we like but doing as we ought.

322

Man is known by the company he keeps; woman by how late she keeps company.

323

It is very easy to find reasons why other people should be patient.

324

Housework is something you do that no one else notices unless you don't do it.

325

The direction we are facing has a lot to do with our destination.

326

No matter how long you nurse a grudge it won't get any better!

327

The richest man in the world is not the one who still has the first dollar he ever earned - it's the man who still has his first friend.

328

The greatest remedy for anger is delay.

329

One pound of learning requires ten pounds of common sense to apply it.

330

To overlook the little things in life is to miss the biggest part of life itself.

331

A friend is a present you give yourself.

332

Habit, if not resisted, soon becomes necessity.

333

The best way to get a day's work done is to work a full day.

334

Few people can stand to hear what they ought to hear.

335

Nearly everyone thinks of sin in terms of what other people are doing.

336

An intelligent girl is one who knows how to refuse a kiss without being deprived of it.

337

A pessimist is a person who is seasick during the entire voyage of life.

338

Gossip is a grapevine that grows only sour grapes.

339

Happiness is best kept when it's given away.

340

You must know you don't know before you can learn.

341

A decision is a choice between alternatives; it is rarely a choice between right and wrong.

342

Hearty laughter is a good way to jog internally without having to go outdoors.

343

Man is the only creature who, when he has all he needs, wants more.

344

The past is a nice place to visit but you wouldn't want to live there.

345

History is 99 percent the achievement of people who never made history.

346

The important thing about your lot in life is whether you use it for building or parking.

347

One things about a computer - before it talks to you, it checks its brain to see if it has anything to say.

348

Instead of putting others in their place, put yourself in their place.

349

History repeats itself because no one listens the first time.

350

Don't think people judge your generosity by the amount of advice you give away.

351

The driver who has "one for the road" is likely to have a highway patrolman for a chaser.

352

Wouldn't it be nice if we could find other things as easily as we find fault?

353

Teachers affect eternity; they can never tell where their influence stops.

354

Be sure brain is engaged before putting mouth in gear.

355

When we are wrong, make us willing to change. And when we are right, make us easy to live with.

356

You can't be enthusiastic and unhappy at the same time.

357

It takes 72 muscles to frown, but only 16 to smile; so, if you are tired, smile.

358

A fanatic is someone that can't change his mind and won't change the subject.

359

The reason why we have two ears and only one mouth is that we may listen more and talk less.

360

The person rowing the boat seldom has time to rock it.

361

Some people's attitudes are like concrete, thoroughly mixed and hard set.

362

Life is like a bank. You can't draw out what you haven't put in but what you have put in, is sure to pay interest.

363

The cost of friendship is only a smile, a friendly handshake and a kindly deed.

364

Frustration is not having anyone to blame but yourself.

365

Don't ask questions if you don't want to hear the answers.

366

An effective boss is one you can work with, not for.

367

There would be a lot fewer pedestrian patients if there were more patient pedestrians.

368

Nature gave you your face, but you have to provide the expression.

369

The object of teaching is to enable the student to get along without his teacher.

370

It's smart to pick your friends - but not to pieces.

371

We cannot avoid growing old, but we can avoid growing cold.

372

It is better to bite your tongue than to let it bite someone else.

373

Opportunity often knocks, but it has never been known to turn the knob and walk in.

374

What they leave IN their children should concern parents more than what they leave TO them.

375

No man ever gets very far pacing the floor.

376

It's not the hours you put in that count, but what you put in the hours.

377

If you're going around in circles, you may be cutting too many corners.

378

Experience is a hard teacher. It tests first and teaches afterward.

379

Every day the world turns over on somebody who had just been sitting on top of it.

380

Snobbery is a confession of one's lack of class.

381

It is said that a chip on the shoulder is about the heaviest load anyone can carry.

382

One way to live a long life is to give up all the things that make you want to live a long life.

383

Contentment is the reward collected by those who feel that what they have is better than what they are missing.

384

If it were not for the doers, the critics would soon be out of business.

385

We spend half our lives answering life's questions and the other half questioning life's answers.

386

Don't talk about your good intentions - do them and others will do the talking.

387

Nothing will be done at all if one waits until he can do it so well that no one can find fault with it.

388

It is not the position but the disposition that makes life worth living.

389

When we really learn to face reality, we will probably realize how well off we were before.

390

All of us want to live a long time, but none of us want to get old.

391

It is just as well that justice is blind; she might not like some of the things that are done in her name.

392

One of the best things to have up your sleeve is a funny bone.

393

The man who removes mountains begins by carrying away small stones.

394

The trouble with burning the candle at both ends is that you always get caught in the middle.

395

You cannot lead anyone further than you have gone yourself.

396

Discontent is the penalty we pay for being ungrateful for what we have.

397

The more we count the blessings we have, the less we crave the luxuries we haven't.

398

You cannot climb the ladder of success with your hands in your pockets.

399

If you find a path with no obstacles, it probably doesn't lead anywhere.

400

Authority makes some people grow - others just swell.

401

Most folks you meet talk too much about the cost of living and too little about the value of life.

402

Tact is the art of building a fire under people without making their blood boil.

403

In the long run, the pessimist may be proved right, but the optimist has a better time on the trip.

404

A smile is something that adds to your face value.

405

Criticism is most effective when it sounds like praise.

406

What sunshine is to flowers, smiles are to humanity.

407

Problems are only opportunities in disguise.

408

Autos are now considered a necessity and children as luxuries.

409

Square dancing is friendship set to music.

410

You can't spend a lifetime just going along for the ride. Sooner or later, you'll have to help pull the wagon.

411

The philosophy of one century is the common sense of the next.

412

God wisely designed the human body so that man can neither pat his own back nor kick himself too easily.

413

To speak without thinking is to shoot without first taking aim.

414

Persons thankful for little things are certain to be the ones with much to be thankful for.

415

If you want to make sure you'll remember your anniversary, just forget it once.

416

Friends are the bacon bits of the salad bar of life.

417

Life is a drawing without an eraser.

418

It sometimes seems easier to do and die than it does to reason why.

419

A born diplomat is someone who remembers your birthday but forgets how many you've had.

420

If you can't have what you want, change your mind.

421

The trouble with the world is that the ignorant are sure of themselves and the intelligent are full of doubt.

422

Patience is a virtue that carries a lot of wait.

423

Don't be afraid to ask dumb questions. They are easier to handle than dumb mistakes.

424

What you are is more important than what you've got.

425

It is the loose ends with which men hang themselves.

426

You never stub your toe unless you're moving forward.

427

The only difference between stumbling blocks and stepping stones is the way we use them.

428

Not everything that is faced can be changed but nothing can be changed until it is faced.

429

You may not always win - but you can always try.

430

A true friend is someone who says nice things about you behind your back.

431

Great men may not have good luck but they can overcome bad luck.

432

A person who says it can't be done is often interrupted by a person who is doing it.

433

Character consists of what you do on the third and fourth tries.

434

People and pins are useless when they lose their heads.

435

Life is like riding a bicycle. You don't fall off unless you stop pedaling.

436

Birthdays tell how long you've been on the road, not how far you've travelled.

437

To be emotionally committed to someone is difficult, but to be all alone is impossible.

438

Rather an honest slap than a false kiss.

439

It is never very clever to solve problems. It is much more clever to avoid problems.

440

You cannot meet the crisis of today, tomorrow.

441

The quieter you become the more you can hear.

442

Success comes in CANS; failure comes in CANT'S.

443

We cannot all perform, but we can all clap.

444

The difference between perserverance and obstinacy is that one comes from a strong will and the other from a strong won't.

445

You can't change the winds of life, but you can adjust your sails to reach your goals.

446

To be efficient do things right. To be effective, do the right things.

447

A prudent man is like a pin - his head prevents him from going too far.

448

It's nice to be important, but it's more important to be nice.

449

You can sometimes catch a terrible chill waiting for somebody else to cover you with glory.

450

When one has a clear conscience, it often means he has a bad memory.

451

He who cannot put his thoughts on ice should never enter a heated dispute.

452

We must choose between Peace on Earth or earth in pieces.

453

Always try to drive so that your license will expire before you do.

454

Character is not made in a crisis - it is only exhibited.

455

There is nothing wrong with the younger generation that the older generation didn't outgrow.

456

The one thing worse than a quitter is the man who is afraid to begin.

457

If you don't learn from your mistakes, there's no sense in making them.

458

Do not resent growing old - many are denied the privilege.

459

Freedom is not the goal, but you need freedom before you can decide what the goal is.

460

You can live on earth only once, but if you live right, once is enough.

461

Footprints in the sands of time are never made by sitting down.

462

Experience is yesterday's answer to today's problems.

463

You can meet friends everywhere, but you cannot meet enemies anywhere - you have to make them.

464

It is easier to keep up than to catch up.

465

The best way to kill time is to work it to death.

466

Half of being smart is knowing what you are dumb at.

467

There are so many substitutes on the market, it's sometimes hard to remember what the original was.

468

All people still smile in the same language.

469

If you're over the hill why not enjoy the view?

470

Kids won't remember if the house was all neat, but they will remember if you read them stories.

471

Right decisions are a result of long experience and long experience is a result of long inexperience.

472

Don't wait for great opportunities. Seize common, everyday ones and make them great.

473

You may out-distance, out-bluff, and out-brag the other drivers, but will you out-live them?

474

Some folks are so busy being good they forget they should be busy doing good.

475

Love talked about can be easily turned aside, but love demonstrated is irrestible.

476

Better than counting your years is to make all your years count.

477

Life is like a grindstone - whether it polishes you up or grinds you down depends on the stuff you're made of.

478

You're not driving your car after you pass 65 miles an hour, you're aiming it.

479

There is nothing so useless as doing things with great energy and efficiency which should not be done at all.

480

The best thing to spend on your children is time.

481

Tell me and I'll forget - teach me and I'll remember - involve me and I'll learn.

482

Identify yourself by accomplishment rather than words.

483

If you would reap praise, you must sow the seeds of gentle words and useful deeds.

484

Sympathy is two hearts tugging at one load.

THE

END

Proofread by "Gene"

INDEX